A THEORY OF EVERYTHING
including
CONSCIOUSNESS AND "GOD"

A THEORY OF EVERYTHING
including
CONSCIOUSNESS AND "GOD"

BILL HARVEY

The Human Effectiveness Institute
Gardiner, New York

Published in 2023 by
The Human Effectiveness Institute

Copyright © 2023 by Bill Harvey

All rights reserved.

No part of this publication may be reproduced or transmitted in any form or by any means, electronic or mechanical, including photocopying, recording, or by any information storage and retrieval system, without permission in writing from the Publisher. Reviewers may quote brief passages.

ISBN: 978-0-918538-19-2 Trade Paperback
978-0-918538-20-8 eBook

Library of Congress Control Number: 2023908401

For information write to:

The Human Effectiveness Institute
12 Amani Drive
Gardiner, NY 12525

HumanEffectivenessInstitute.org

Cover art (Sacred Geometry 9) and design by
by Endre Balogh / EndresArt.com

Editing, book design and typesetting by Yana Lambert

Dedicated to
Albert Einstein & John Wheeler

CONTENTS

Preface	ix
1 Introduction	**1**
Inner Life	1
The Benefits New Science Can Bring	6
What Exactly Is Needed	7
A Role for Individuals	8
2 Unified Field Theory—	**9**
Theories of Everything	
"Bits Before Its"	12
The Participatory Anthropic Principle	12
What Changed During the History of Science	13
Materialistic Accidentalism	14
The Meaning of Life	15
Separating God from Organized Religion	17
How Can We Reconcile "God" with Science?	18
Defending the Idea that Consciousness Came First	20
A Possibility to Consider	21
The Better of Two Bootstraps	23

What would you do if you were the first self?	23
Never a Beginning	24

3 A Simulation of What the Most Recent "Beginning" Might Have Been Like — 25

4 How Could One Imagination Create a Universe of So Many Beings? — 29

5 The Singularity — 35

6 Extrasensory Perception — 43

7 Suffering — 49

8 Experimentation for Individuals — 53

Experiment 1	53
Experiment 2	58
Experiment 3	60
Experiment 4	61
Experiment 5	62
Experiment 6	62

About the Author	64
Further Reading	67

PREFACE

It is long past time to discuss the ultimate questions such as "why are we here?"—those questions we set aside because things were interesting enough in the world of the senses and everyday pleasures to keep us distracted from the deeper questions in life.

Now, when things have gotten rough and the existential challenges are many, we remember that our homework is not done. We *need* the answers to *why* we are keeping up our daily routines, in order to have the intestinal fortitude to stand up and do whatever it takes to overcome the new challenges facing civilization today.

I wrote this short book mainly for physicists but also for everyone. I'd like physicists to accept the scientific possibility of something very much like "God", and to prioritize the subject. This book explains why. I hope you enjoy it and feel what Freud called "the Oceanic feeling".

φ

I would also like to express my gratitude to Yana Lambert and Lalita Harvey for their invaluable ongoing editing guidance, and to the multitalented artist Endre Balogh for his exquisite art gracing the cover and his simply elegant rendering of the Sri Yantra which opens each chapter.

—Bill Harvey
Gardiner, New York
March 2023

CHAPTER 1
INTRODUCTION

Inner Life

From the start, many of us could not shake off a strong compulsion to know more about who we are and why we are here.

Early aspirants to self-knowledge attained high levels of sophistication in contemplating the self, remembering key ideas from generation to generation as lyrics in songs or lines in poetry or symbolically in art, especially before the advent of the written and then printed word. Spirit and mind were regarded as the same thing early in human development, which placed psychology and spirituality in the same category. Most of these key ideas being passed from one generation to the next relate to the Oneness, the interconnectedness, of all things. The *Rig Veda*, *Egyptian Book of the Dead*, *Torah*, and *Tao Te Ching* are classic examples.

It is logical for science to consider this as evidence of a long-running human intuition that there is a far greater intelligence overseeing the universe, and that we are all parts of that intelligence. Just because almost everyone at some point experiences such an intuition does not prove that there is anything scientific about it.

However, it is reasonable to keep an open mind. Science is supposed to do that all the time anyway; the only exception appears to be in this one subject domain, "God", as if a flinch reaction to having been "deluded" by the "God" concept for so many millennia. Hopefully, most scientists do have open minds, because—as we aim to prove—this subject needs more attention, *stat*.

The "inner life" tradition—pondering the ultimate questions, studying the self—has continued alongside every step forward we have made in science and technology. In the world of 2023, hundreds of millions of people practice meditation and yoga worldwide, taking part in at least a basic course of self-examination and thought experiments into what the world might be. The most advanced thinkers of every era gave us (sometimes only through the word-of-mouth of their followers) the findings of their own inner explorations.

For 6000 years of written history—describing as far back as 10,000 BC based upon oral "records"—the ideas of materialism and spiritualism had always been of nearly equal importance in explaining people's behavior. That healthy balance took an odd turn a few hundred years ago, starting first with a shrinkage in the average person's own spirituality, and then in the outward self-identification people give in surveys. This last drop took place only in the past 20 years, in which self-proclaimed spirituality (usually an exaggeration of true spirituality) dropped in the U.S., for example, from 90% to 70%[1].

President Eisenhower's speech as he left office described Americans as "a free and religious people". The absurdity of saying that in 2023 seems obvious. The daily news paints all of humanity as running amok. People invoke God to sanctify deeds of extreme hostility, while

---

(1) https://www.pewresearch.org/religion/2022/09/13/how-u-s-religious-composition-has-changed-in-recent-decades/

politicians use God as a wedge issue, pitting neighbor against neighbor. Perhaps they're still thinking of the old vengeful images of God we first developed while hiding in caves.

The "hip" thing to say since around 1970 is a bit like "I believe in something—some kind of force—it may be nothing like us—it may not care about us—but it is there." This is the way to convey that you are a spiritual, and therefore a good, person—another concept of God.

The deists who founded the USA believed that whatever God was, "He" was supremely benevolent and had our best interests at heart. And they believed that the teachings of Judeo-Christianity and Islam well describe what God would want of us, for our own good. We trust in God. Even though we don't know what "He" is.

Except that nowadays, even with the social pressure to go along with the herd, a third of us boldly state that we don't trust God or even acknowledge that "He" exists.

Eisenhower was not alone in his association of spirituality with the economic, social, and political success of a nation. He *did* trust God.

A plausible social science theorem would be that the degree of civility in a culture is predicted by its degree of trust in God. If we trust that a benevolent intelligence protects us all, we are less likely to overreact to inconsiderate behavior.

If we are to trust God in this modern cynical age where society seems to be mostly ruled by violent mood swings, we cannot avoid discussing the nature of God within a scientific framework.

If God cannot be reconciled within science, the game is over. The denouement shall not be pretty. The cards are stacked against us: Hedonism, materialistic accidentalism, cynicism, selfishness, egotistical self-centeredness, scarcity of wealth (only a few have it), power-mongering, hatred of "the other", racism, misogyny, xenophobia, fear, weapons of mass destruction, collapsing environment, dying species, fiat money, superficial education systems, unmoderated platforms of social mass communication—all of which are the self-destructive negative forms of creative expression in the absence of spirituality.

Civilizations have been here before; we know that such forces have been capable of bringing down great empires, disposing millions of people to have to start all over again.

The Benefits New Science Can Bring

If science concludes that there is no scientific basis for ruling out intelligence in the universe itself, the ensuing shift in the bedrock assumptions of the culture is likely to gradually bring about the strongest upsurge in positive emotion in his/herstory. For the same sorts of reasons as "The Good News" brought by Jesus was perceived as cause for happiness by his followers then and now.

Science's acknowledgement of possibilities very similar to the existence of God—a conscious, intelligent universe, with a logical desire to protect itself and its parts, functionally the equivalent of benevolence—makes each of us important again since we were important enough to have been created by something like God. Life is important and has great meaning, not to be treated frivolously or callously.

Hypotheses: Better behavior will steadily take over. Working together in a friendly way will become the norm. People doing things they enjoy shall be the employment principle, inspiration and creativity and compromise shall all blossom, and love will flower.

What Exactly Is Needed

These are the upside benefits to humanity if science proclaims that there is no scientific basis for ruling out a benevolent God whom we are right to trust. Such a new position would encourage us to take a fresh look through the lens of science at how consciousness might be related to matter and energy, for that is the experimental direction to determining what "God" there really might be.

A "fresh look" means *not* again and again solving for the matter-energy quantum brain phenomena which generate the epiphenomenon of consciousness. Instead, science should be studying and explaining why we experience consciousness, why and how Flow state happens, why there are other states of consciousness and powers of mind that come and go.

We need theoretical scientists to focus on all of the possibilities for how the universe came to be, without any presuppositions.

For example, could Wheeler's first cause "quantum foam of probabilities" exist within a consciousness? Wheeler was a great coiner of words and phrases, and was the first to use the term "quantum foam", although he was describing the implications of the ideas of Werner Heisenberg in that coinage.

The standard interpretation of quantum theory states that even in nothingness, something is going on, and that something consists of probability waves which spontaneously morph into virtual particles which appear and disappear.

To us, that sounds more like what goes on in a consciousness, not like what goes on in a material world. Why not design experiments by which to test that hypothesis?

A Role for Individuals

A second experimental path, which individuals could take, is described in the last chapter of this book: how an individual can test various ways of using his or her own consciousness, and observe what the results might be. These results may not be of use to science, but could turn out to give people useful new tools and possibly even a profound spiritual feeling.

CHAPTER 2
UNIFIED FIELD THEORY— THEORIES OF EVERYTHING

Einstein died before completing his Unified Field Theory, which would have explained how gravitation, electromagnetism, and the strong and weak nuclear forces fit together and why each of those four forces exist at all.

He did not explicitly state any intention to include consciousness with the four physical forces ("physical" meaning matter *and* energy). However, without connecting the dots for us, his thought experiments which led to his relativity theories always included an observer. Somehow, he needed to use an observer in order to describe how reality works. He did not use the word "consciousness" but it is implicit in the word "observer".

Einstein was intuitively certain that all scientific discoveries, which revealed to him beautiful complexities beneath the appearance of things, proved that there was an incredible intelligence behind the universe. To him it was highly unlikely that everything came together to form this universe completely by the accidental crashing of matter and energy.

> *Certain it is that a conviction, akin to religious feeling, of the rationality or intelligibility of the world lies behind all scientific work of a higher order... This firm belief, a belief bound up with deep feeling, in a superior mind that reveals itself in the world of experience, represents my conception of God.*
> —Albert Einstein[1]

> *The God Spinoza revered is my God, too: I meet Him every day in the harmonious laws which govern the universe.*
> —William Hermanns[2]

---

(1) *Ideas and Opinions*, "On Scientific Truth" (New York: Crown Publishers, 1954), p 261.
(2) *Einstein and the Poet* (Brookline, MA: Branden Press, 1983), p. 9.

*I shall never believe that God plays dice
with the world.*
—Albert Einstein[3]

*In essence, my religion consists of a humble
admiration for this illimitable superior
spirit that reveals itself in the slight details
that we are able to perceive with our frail
and feeble minds.*
—Albert Einstein[4]

John Wheeler, who had been highly influenced by Einstein,[5] went further than his mentor in drawing connections between consciousness and matter-energy. His theories of consciousness within a quantum physics framework evolved in two phases.

---

(3) Philipp Frank, *Einstein, His Life and Times* (New York: Knopf, 1947).

(4) April 24, 1929 in response to the question of New York's Rabbi Herbert S. Goldstein: "Do you believe in God?"

(5) Contrary to popular belief, while Wheeler and Einstein collaborated on a unified field theory of the physical forces of nature, they did not collaborate on the building of the first atom bomb. Einstein's famous equation E=mc2 explains the energy released in an atomic bomb but doesn't explain how to build one. To quote Einstein, "I do not consider myself the father of the release of atomic energy. My part in it was quite indirect... I believed only that release was theoretically possible. It became practical through the accidental discovery of chain reactions, and this was not something I could have predicted." (*The Atlantic*, November 1945, "Einstein on the Atomic Bomb"). Wheeler in his autobiography writes extensively about his work in the Manhattan Project. Although his role was limited to the extension of nuclear theory, he had a personal reason for wanting to accelerate the bomb project so as to save lives: his younger brother had been killed in action during the Allied invasion of Italy.

"Bits Before Its"

Wheeler concluded that the substrate underpinning matter-energy—asteroids, people, dogs, mountains ("Its")—was preceded by encoded information ("Bits"), which was the blueprint for the It and its cause of existence.

This is eerily similar to the words of the Bible in Genesis, in which the matter-energy universe was created by "the Word". A word is a form of encoded information.

The Participatory Anthropic Principle

In Wheeler's ultimate view of reality, matter-energy preceded the existence of consciousness, and existed as probability waves rather than as concrete "Its"— not yet what we apprehend as matter and energy, which only became what we see and feel after our consciousness came into existence.

Wheeler did not comment on whether the universe was random or guided in its development of consciousness, but he did state that it was as if the universe somehow knew it needed to develop consciousness in order to collapse probability waves into the universe that we are able to behold.

The Question is what is The Question?
Is it all a Magic Show?
Is Reality an Illusion?
What is the framework of The Machine?
Darwin's Puzzle: Natural Selection?
Where does Space-Time come from?
Is there any answer except that it comes from consciousness?
What is Out There?
T'is Ourselves?
Or, is IT all just a Magic Show?
Einstein told me:
"If you would learn, teach!"
—John Wheeler[6]

What Changed During the History of Science

Reading about Wheeler and making inferences by reading between the lines, it seems that Wheeler did not want to risk his reputation by defying what has gradually become an unwritten convention of modern science: assume that the human mind's intuition of (and/or belief in) an intelligence as the source of the universe is "magical thinking", "superstition", and "anti-scientific".

---
(6) Speaking at the American Physical Society, Philadelphia (April 2003)

It has not always been that way. Going back to the earliest scientists (originally called "natural philosophers") there was an easy coexistence with the idea named "God", which goes back much further than written language. For Socrates, Plato and Aristotle, Galileo, Newton, Leibniz—the list goes on and on—there was never a binary schism forcing them to choose between spirituality and science. That hard line has been drawn in just the last few centuries. We take this dichotomy for granted due to our cultural conditioning. But the dichotomy is itself a theory, not a proven fact: no scientific proof exists that rules out the existence of an intelligence in the universe itself.

Materialistic Accidentalism

The denial of "God" (intelligence in the universe itself) became a fashionable style among scientists. It was fairly easy to convince many people to give up on that old-fashioned notion of God. The omnipresent wars and injustices going on around us seemed supportive of the idea that all of reality was an accident in the first place, leaving us to deal with the "dog eat dog" bar brawl however we can.

Deeper thinkers rationalized the ability of crashing matter and energy to build complex and self-reproducing structures by accident, saying, "In infinite time, everything has to occur," like a million monkeys playing with typewriters eventually writing "Hamlet". But we have never observed crashing waves on a beach building a turreted, arch-windowed sandcastle. Nor does probability theory inherently contain any mathematics that would require all possibilities to come to pass.

Nevertheless, people absorb the biased information they receive, and each individual makes a worldview out of it. At the present time (2023) the human race collectively appears to largely pay lip service to the idea of "God" if they live in a place where this is the tradition, or they call themselves "atheists" if their personal community (e.g., most scientists today) has greater respect for that side of the dichotomy.

The Meaning of Life

Each human being can choose what their purpose shall be in life. This choice can be made independently of the choice of what to think about the question of "God". However, there is a covariance between these internal personal decisions. People who actually feel that God exists tend to choose more noble purposes in life, whereas people who are certain that God is a fiction tend to be motivated by money, power, sex, fame, and being treated with respect more than anything else.

There are of course notable exceptions, humanists who are altruistically motivated without need to base that on any ontology. Those humanists are sometimes scientists such as Einstein, who worshipped the brilliant beauty he saw in obviously intelligent Nature, which he saw in the intelligent universe itself. Others express themselves in similar ways without spelling out their vision of intelligence or consciousness in the universe itself:

> *My atheism, like that of Spinoza, is true piety towards the universe and denies only gods fashioned by men in their own image, to be servants of their human interests.*
> —George Santayana[7]

> *Human decency is not derived from religion. It precedes it.*
> — Christopher Hitchens[8]

With the U.S. moving from the aforementioned 90% to 70% claiming to be spiritual in the last 20 years, and knowing that those truly "behaving spiritually" are a subset of even the 70%, the motivations we might refer to as "baser" (less noble) are on the rise. We've already seen the consequences of this decline in spiritual identification and such consequences are not

---

(7) *Soliloquies in England, and Later Soliloquies,* "On My Friendly Critics" (1922)
(8) *God Is Not Great: How Religion Poisons Everything* (New York: Twelve Books, 2007)

going away just because individual leaders are being voted out of office, perhaps on the contrary.

Having an open mind rather than a prejudged bias about whether the universe is intelligent would appear to have some relation to the existential challenges humanity now faces.

Separating God from Organized Religion

Organized religion has done its fair share of good though has done more than its share of harm through the ages. The cultural cancelling of "God" in recent history has apparently subconsciously undermined even clerics, causing some of them to behave as if they too are motivated by money, power, sex, fame, and a craving for respect, from which we might infer that their mental/emotional grip on the concept of God has been loosened. The Bible recounts that even thousands of years ago, well before modern science, clerics could similarly go astray. Today's atmosphere of materialistic accidentalism in science only feeds such derelictions within religious organizations.

The suggestion to separate "God" from organized religion is offered in this sense: many people jump to the hasty closure of throwing out the baby, God, with the bathwater, organized religion. In considering that the universe may itself be intelligent and may have had a creative role in bringing about our conscious existence, we need to peel away that objective reconsideration from irrelevant side issues like the good versus harm done by organized religions throughout history. One has nothing to do with the other in terms of logical rational thought; they are separate questions, and must not be muddled together.

How Can We Reconcile "God" with Science?

Because of the habits of the human mind, especially in a culture in which humans have created more complexity than our minds can easily handle, it may initially be useful to temporarily set aside the word "God" and speak only of a conscious, intelligent universe.

It is far easier for today's human mind to objectively consider the possibility that something as big and as filled with inanimate objects as the universe could itself have intelligence, than to discuss a word so saddled with millennia of baggage associations. The word itself unleashes emotions, chemicals in the body, muscular reactions, imagery, feelings beyond description. Let's park the word and continue the investigation of where we are at this crossroads of life and self-extermination, and how our current exigency relates to our thinking and ways of being.

Wheeler again has theorized that consciousness is a real thing and has vast importance in the scheme of things in this universe in which we live. Consciousness, according to Wheeler, transforms a universe of probabilities into a world of tangible matter and energy events.

Science has not rejected Wheeler's ideas. It has largely ignored them. That is, science has ignored those ideas of Wheeler's that have a bearing on the existence of consciousness as an essential aspect of the universe. Science has certainly not ignored his other ideas about black holes, nuclear fission, thermonuclear fusion, quantum foam, or wormholes.

Given the scientific community's respect for Wheeler and the non-rejection of his theories about consciousness and the universe, it should not be too difficult for scientists to accept the possibility that Wheeler may have been right about everything, except perhaps the sequence of early universe events.

Our theory is that Wheeler was incorrect about consciousness coming after the beginning of the universe. It makes more sense that before matter and energy, there was consciousness, which compelled matter and energy to come into existence.

Defending the Idea that Consciousness Came First

Any cosmological theory faces the challenge of explaining why there is a universe at all. Logic suggests that nothing should ever have existed. Something cannot come from nothing. Therefore, there must always have been something.

In scientific thought today, it is Wheeler's quantum foam of probabilities that was always there. Then the big bang came from that, and eventually crashing matter and energy led to self-reproducing complex structures accidentally, and those eventually became life, and life eventually brought forth brains, and brains generated consciousness.

Doesn't this picture seem overly optimistic about what can come about accidentally?

Not to mention the question of where the quantum foam of probabilities came from.

Science has made it a tradition to dodge these questions of how things started.

Glimmers of light appear from time to time. Today most physicists acknowledge that "the hard question" is how to incorporate consciousness properly in the unified theory of everything. This is the direction from which science can begin to theorize about the start of the universe.

A Possibility to Consider

Let's imagine what it could have been like before the existence of what we experience as the universe.

Imagine total nothingness. No quantum foam probabilities, no anything. Just endless nothingness.

Imagine that after the passage of unimaginable amounts of time, the nothingness realizes itself as something noticing a persistent experience of nothingness: the Noticer.

The time that has passed is merely the subjective experience of the nothingness that has always existed in the mind of the Noticer.

"The nothingness has always existed, it exists right now, and will probably go on existing forever," might have been the first intuition of the Noticer.

"I AM THAT nothingness" might have been the next intuition the Noticer had.

"I am the Nothing's imagination", might have been the third intuition.

That Consciousness could have continued to think and found it to be more fun than just watching nothing happen forever.

Why did we just slip in "Noticer" and "Consciousness" with initial caps? If we are considering a scientific proposition regarding a theoretical consciousness of the universe itself, it seems proper respect to use initial caps.

Does it follow that all of the connotations of "God" are to be assumed of the consciousness of the universe? Not necessarily.

What we are suggesting is that, if nothing else, it is simpler to assume that a persistent experience of nothingness could lead to the experiencer realizing that it exists as an observer—simpler than to imagine that a quantum foam of probabilities existed, exploded, and things slammed against each other until this world we see around us in lightyears in all directions came to be in all its wondrous complexity, and eventually created consciousness, the ability to perceive oneself as a persistent entity which experiences things.

The Better of Two Bootstraps

The standard model at the moment is that a complex physical form evolved from random collisions we call The Replicator Molecule, and thus life came to exist.

The model we present here is similar in that it starts with random information bits representing nothingness, assembling a self-referential viewpoint, a permanent memory-creating self.

One could argue that it is less implausible to envision random information becoming a self-organizing system than it is to envision random collisions of matter-energy building any complex physical thing let alone one that is also a factory for others of its kind.

What would you do if *you* were the first self?

There you are, you just realized that you exist, and you are alone amidst nothingness.

You might think and think and think and at some point, come to the conclusion that you and imagination are one and the same.

This might lead to experimentation as to how far you could go just by imagining things. How intensely could you visualize something else besides nothingness? How real could you make your imaginings seem to you?

After all, once having become consciously self-aware, were you going to simply accept nothingness as your way of life forever? Or would you want to at least try for something else?

What else was there to do but to explore one's own capabilities? How far could imagination be pushed?

Never a Beginning

Although the better of two bootstraps is appealing, a simpler theory is that it has always been this way. There never was a beginning.

Our present theory is that time itself is not intrinsic to the One Consciousness, who has the computing power to experience all time at once. Time is part of the imaginary world the One Consciousness creates and inhabits through its avatars.

The expanding universe since the Big Bang suggests a cycle similar to an inbreath alternating with an outbreath, with all of creation sucked back into the Creator for what might be a sleep cycle, followed by a reawakening expansion.

CHAPTER 3
A SIMULATION OF WHAT THE MOST RECENT "BEGINNING" MIGHT HAVE BEEN LIKE

Imagine that nothingness could discover its own consciousness. The effect would be one of awakening to a consciousness, just like we all have, but at an infinitely higher level of power—an intelligent consciousness that experiments to determine what powers it might have, and realizes that it can instantly materialize its own imaginings, and that it can inhabit an infinite number of avatars of itself.

Each of us is conscious, therefore we can imagine what it would be like to experience nothing forever and then realize that something is persistent in the situation—a self—one's own self.

If you clear your mind and meditate on what it would be like to wake up with amazing powers but nobody to play with, you are likely to start thinking along the lines of the universe's consciousness—using your imagination to create avatars, and then inhabiting at least some of them in such a way as to not remember that you are The One Self in control of everything and enjoying all of it.

These purposely anthropomorphic projections of what the universe might feel if it is conscious are aimed to break down the barriers that restrict our ability to consider different interpretations of the evidence. If we widen our epistemology on a trial basis, we see that a scientific explanation could allow for the universe being conscious and intelligent.

Each part of me in an avatar—my created being—can forget who I am, but I will also stay awake in my full identity at the same time. I'll see the view from each avatar's lens and I'll also see my own view of everything all at once. Fortunately I seem to have infinite attention span. I guess that's because all of this is just taking place in my own imagination anyway.

The parts of me in each created being that don't remember they are me will be like temporarily Lost Lambs, but eventually they will wake up and remember that they are me.

I'll give each one as long as he or she needs. Some may need to change worn-out bodies before they wake up. No problem.

Each one, of course, will have free will, just like me. They will have no real perspective to begin with—no memories of where they came from—so they will proceed by trial and error, and learn along the way.

What a game!

CHAPTER 4

HOW COULD ONE IMAGINATION CREATE A UNIVERSE OF SO MANY BEINGS?

Consciousness itself is a form of information processing. The closest analogy would be to call consciousness an energy computer—a computer made out of energy.

The consciousness level of a human being could never do all the things that the theoretical First Self has demonstrated in the universe we experience. We are limited to some degree by the physical constraints of our brain and nervous system. However, even were we to be separated entirely from our physical body and discover that our consciousness still exists—if that is the case—it appears unlikely that we would at once have the functionality of mind to encompass the view into the inner perspective of all conscious beings in the universe simultaneously. Why is this so?

Human beings can partition their minds only to some extent. In playing the piano, for example, the left hand and the right hand are doing different things at the same time. And creatures who are thought to be less intelligent than humans, such as octopi, have a brain in each tentacle, suggesting that each arm-leg might have its own sense of self which the whole creature is also able to experience through its many brains.

But living through each mind in the universe is a far more daunting venture. As we know our own capabilities today, we would not be prepared for the experience of suddenly being able to see out the eyes and other sensory equipment of every mind in the universe. in that situation, we would most likely experience complete overload—a chaos of unintegrated fleeting thoughts, feelings, images and impulses. It would likely be quite unpleasant.

What accounts for this vast difference in mental capacity? It would seem to come down to the amount of energy accessible by the mind in question. The average person consumes and expends up to a few thousand calories worth of energy each day. The total universe production of energy per human twenty-four hour day is almost infinitely larger. The First Self must have always had energy on that scale. Being able to imagine many creatures into existence and then enter them to live through them—with or without memory of being The First Self while in the creature—and to be able to enjoy the multiplicity of lives all at once suggests an infinite or virtually infinite attention span.

The First Self may have created beings vastly more mentally capable than we are, we just haven't met them yet. Or their existence has yet to have been documented as fact.

It could also be that living through creatures with certain degrees of mental constraint— and doing so with no memory of actually being The One Self—enables The One Self to enjoy more powerful experiences with zero sense of them being just imaginary. Whereas from His/

Her original perspective, it would seem to be all too obvious that all of it is just imaginary.

Speculating further, the personas created by His/Her created beings experiencing many lives sequentially (if, as some believe, that is what is going on) could create complex personalities that The One Self could enjoy as companions or "apprentices" over vast periods of time.

The One Self could help these avatars learn how to handle more simultaneous inputs and challenging ethical questions that would make it less risky to allow them to be born into beings of greater power.

At the end of a long course of instruction and demonstration, a self that has accumulated wisdom and understanding over many lifetimes in an ascending array of levels of being (powers of mind) could be reabsorbed into The One Self as a conscious sub-self, a personality aspect imbued with the deep learning of experience.

If this is the sort of game the Universe is playing, even at our limited human level we can understand and appreciate the Universe's motives. We are witnessing consciousness at play. But simply the fact that it is playful does not mean that it is mere entertainment. The

Universe is its own art form. It is aesthetically beautiful.

φ

The theory that consciousness came first and created matter-energy—The Theory of the Conscious Universe—is very close to the theories put forth by all Eastern spiritual traditions such as Hinduism, Taoism and Buddhism, and virtually identical to Kashmir Shaivism—except that the language being used here is closer to the language of Western science. The intent of this book is to persuade the reader to open-mindedly consider the possibility—and perhaps even probability—that something like the scenarios painted here portray the real truth about what the universe is.

What about Western spiritual traditions? To what extent does The Theory of the Conscious Universe square with Judaism, Christianity, Islam, and Baha'i?

All of these Western spiritual traditions stem from the founder's experiences of communicating directly with The One Self and/or with beings more powerful than ourselves who serve The One Self.

The founder in each case teaches that consciousness continues after the death of the body.

And the common main message is to do kindness continuously, and to seek to do what The One Self (whom they call God or Allah) teaches.

In certain branches including Jewish Kabbalists, Christian Quakers, and Islam in general, allowing The One Self to take over control of one's body and mind is a specific teaching.

In at least these four specific areas—receiving guidance from above, consciousness survival after death, the importance of continuous kindness, and surrendering to control by The One Self—the Theory of the Conscious Universe lines up very well with spiritual traditions on Earth.

CHAPTER 5
THE SINGULARITY

The moment in which The One Self becomes self-aware needs to be explored. Why did this singularity occur? Why had it never occurred before?

And what *was* it? If all that had existed was nothingness, something in the nothingness had to have constituted possibility.

Picture this in two different ways, with and without Time:

1. The One Self has always existed and will always exist and is the only thing that exists. It has a waking/sleeping cycle. Every time it wakes up there is no memory at first. Perhaps there is a Big Bang every time The Self wakes up. The expanding universe reaches a certain point then begins to contract back into itself until it enters the sleep cycle. As if the Universe is breathing. In multiverse universe-sheaf[1] imagery, each Universe may manifest this differently.
2. The other way to visualize it is that all time exists at once, and the multiverse effect is compounded by different free will decisions in each starting universe.

Perhaps both views are true and co-exist, depending on how The One Self looks at it. All of this is for the enjoyment of The One Self. With infinite or virtually infinite computing power, The One Self could easily see itself both ways at once, with Time sequence or all-happening-at-once.

The design enables The One Self to pay special attention to minute details within His-Her-Self. Each of us is a special adventure-comedy-romance-drama story unfolding within the

---

(1) *The multiverse idea envisions a sheaf of different universes existing side by side with each other, with each universe branching out to explore the implications of going in one direction or another at crucial seed moments.

grand scheme of things, and what we experience as our "self" is actually The One Self.

How can that be?

Picture the singularity which appeared out of nothingness as a spark. That spark might be likened to a subatomic particle. However, we posit it to have characteristics not normally associated with leptons, mesons, and bosons. Let's call it a "psion", meaning that it has access to "psychic powers" as broadly defined to include the basic psychic power called "consciousness". Consciousness itself is a psychic power. The most distinguishing characteristic that a psion has is its sustained ability to know itself as an experiencer, and even in the absence of stimuli is able to experience thinking, feeling, envisioning, imagining, intuiting, intending, planning, anticipating…

It's acknowledged that these are amazing powers in themselves, especially from the viewpoint of the moment in time just before the first psion existed. ("Extrasensory perception" is addressed in the next chapter.)

Once there are multiple psions, they can interact with one another, providing stimuli one to the other.

Leibniz had a similar thought and his name for the conscious entities was "monads".

In order to fit within current science, psions would either have to be even smaller components than neutrinos, or they would have to be neutrinos.

Technically, in theory, when The One Self creates an offspring of The Original Psion, The One Self teleports a copy of itself with certain functionalities withheld, to a place in spacetime, where that creature begins a series of lives, moving up the evolutionary chain to higher and higher being states, life to life, depending on its speed of learning.

Interestingly, this bears a resemblance to one of Wheeler's theories, that all electrons are actually the same electron at different stages in its life. That Wheeler theory also has a nexus to Plato and his concept of there being a higher level of Universe where the ideal images of archetypal things exist. Consilience of ideas across thinkers suggests evidence of possible if not probable truth.

Although we do not know yet if this is the way reality really is, so far we can claim a degree of agreement between our theory and the existing orthodox standard theory of physics today. Psions become part of the family of subatomic particles/wavicles. Consciousness is the field created around the particular psion. Each psion is actually the same singular spark of experiencing that noticed itself at the beginning of the Universe.

This theory explains why consciousness (the observer) is so important in Einstein's and Wheeler's cosmologies. Wheeler indicated that matter-energy by itself can only have a shadowy existence until consciousness beholds that matter-energy. We submit this is because the entirety of existence is there to be enjoyed, that is its purpose from the beginning. To be enjoyed by the seasoned experiencer, The Original Self, and by each of His/Her offspring.

The Universe which contained nothingness now also contains the singularity—the original spark of imagination which bootstrapped itself into a fullness of Universe populated by progeny each inhabited by Itself. The degree of Its own functionality transferred into each spinoff differs, with human-level psions having uniqueness, free will, animation, reproduction, agency and other functionalities, which subatomic particles appear not to share. Yet even these particles are psions—they are made out of the only substance that exists, consciousness, and thus they are elemental particles/wavicles of consciousness, manifesting as matter-energy. Although it is all consciousness, some of it appears as matter-energy and some of it, such as the thoughts in your head right now, are invisible.

John Wheeler also happened to create the first formula by which light can be bound into matter. In Jewish cosmology and in the Bible, the first matter-energy created by the first point of self-awareness *is light*. Everything else in the world of matter-energy is a modulation of that original light.

At the human level of consciousness, we not only *are* consciousness (also manifesting physical bodies exhibiting some degree of energy) but we also, through imagination and follow-through, can turn our imaginings into matter-energy forms which other human-level psions can appreciate.

Carl Jung was one of the first scientists to postulate the possible existence of a collective unconscious, where all human minds touch one another as if in a pool. Our theory is similar except that the collective consciousness contains everything in the Universe.

Bishop Berkeley had theories similar to the one we present here. He also described a way of being in which one sees oneself as being the only thing that verifiably exists, with everything else imaginary. This is called solipsism. Our theory would then be a species of universal solipsism, where even one's own self is imaginary, and the Imaginer is the secret identity of all psions. The Imaginer is the One Self inhabiting us all.

The functionalities which we describe in the process of The One Self building the Universe of beings and things call to mind familiar phenomena we observe in the natural world. Amoebas split off a part of themselves to create progeny; The One Self does the same. As mentioned earlier, octopi by their multiple brains are equipped to experience multiple selves, one for each tentacle, and the same could be true of other creatures here and on other planets. Some species such as ants, bees and termites may experience groupminds, and watching a flight of birds or a school of fish move as one, suggests that such a phenomenon could be an on/off option available for more evolved species.

These known phenomena demonstrate that consciousness is even more complicated than it seems, and that under certain conditions, psions may be able to share their consciousness. This is the subject of the next chapter.

CHAPTER 6
EXTRASENSORY PERCEPTION

The theory proposed in this book is that there was/is an Original Consciousness that corresponds to, and may be the reality behind, our concept of God. That Original Self has, in our theory at least, created an "arcade Universe" in which That Self may play, "as if" with friends, who are really subsets or spinoffs of The Original Self.

There is no reason to suppose that The One Self would have some built-in barrier preventing the mixing of one person's consciousness with that of another person. Certainly if The

One Self has created all these galaxies, species, and all the incredible complexities science is slowly uncovering, the entanglement of consciousness we call extrasensory perception (ESP) should not be ruled out as a real (scientific) possibility.

The evidence already on the table for the statistical proof of ESP includes famous meta-analyses performed under the direction of Charles Tart.

Further corroboration is found in the existence of rare individuals who appear to have received information from what could be above the level of the human race. These individuals include the Hebrew prophets, the authors of the Vedas, Abraham, Moses, Buddha, Jesus, Mohammed, Mirza Husayn 'Ali Nuri (Baha'u'llah), and many others, possibly including Einstein and Wheeler. This form of ESP is traditionally called inspiration, the receipt of knowledge from above or beyond.

Within the framework of our theory, there is no reason to limit the likelihood of receiving guidance directly from The One Self—since we are each one unique combination of pieces of The One Self. We are also *equal* to The One Self in our awareness of *being a self* (we are the spark of experiencing), though not equal in other functionalities, at least in our current incarnation.

There are many other powers of consciousness that tend to be denied or underutilized in our current version of civilization. Hunches are sometimes accurate precognitions. Sometimes we *are* reading each other's mind. Sometimes we think of a person at the same moment they think of us and we both find that out when one calls or writes the other by material means.

We also experience altered or peak states of consciousness, with or without chemical stimuli. One such state is named the Flow state by Mihaly Csikszentmihalyi, which athletes call the Zone and performers call being "on" — the state in which one is an engaged observer watching oneself perfectly execute the actions intended as if the actions are doing themselves, or as if The One Self has taken control of one's actions.

There are also Out-of-Body Experiences and remote sensing and telekinesis (affecting the motion of other matter-energy objects other than oneself without causing any physical contact with the moved object). All of these are different ways that consciousness sometimes operates.

I personally have experienced most of these phenomena, most of them rarely, some daily.

One of the most useful forms of these peak experiences is the Observer state, a state of relative indifference to outcomes, which may be the prerequisite state to Flow state.

Synchronicity is experienced so often that it appears too unlikely, every time it occurs, to be explained away as mere coincidence. The mindset of asking oneself what the Universe is teaching us by a given synchronicity often yields insights which are of pragmatic value. This tends to cause one to consider that an intentional process is behind these phenomena, a teleology (purpose-driven process) that runs through the whole Universe from its inception through each moment of time in each psion (the aforementioned subatomic particle of consciousness).

Given the widespread denouncement of ESP, one wonders if there were to be no such taboo, would many if not most of us quickly realize for ourselves that these powers exist and can be cultivated?

The receiving of guidance from invisible benefactors is reported frequently not only in the Bible but throughout mythology. There could be a scientific basis for many of these accounts, especially those that changed the world, such as the inspirations received by founders of religions and prime movers behind scientific revolutions. In my youth I used the term "Noia" to mean "The suspicion that someone is out to do you good". Having a worldview that meshes with a propensity to spot cues from helpful invisible sources is beneficial, even if it rests upon a theory which may not yet be perfectly accurate.

Hypothesis: Making scientific peace with God and ourselves as kindred consciousness will result in a more open-minded, less denial-driven, more-inclusive approach to discovering the functionalities within our own personal consciousness, including the power of prayer and the ability to engage in direct personal communication with the essence of yourself, the Creator of the Universe.

Although this is only a theory, the same is true of the theory of Materialistic Accidentalism. The question is, which is closer to the truth? Our theory explains and integrates many phenomena that are neither explained nor allowed for in Materialistic Accidentalism, from ESP phenomena to the brilliant advice given by certain sages that molded our His/Herstory and led to major turning points in our collective life. It is consistent with the model of the connectedness of all things within a single consciousness.

CHAPTER 7
SUFFERING

The Universe is its own art form. It is aesthetically beautiful.

Everywhere one looks one can find beauty — except for the apparent cruelty we see in human life and as we surveil the fight for survival across all the species.

Why would The One Self put itself through such torture, since it is The One Self inside each of those creatures being shot at or waterboarded, electrocuted, stretched on the medieval rack, or eaten alive?

It's the price The One Self is willing to pay for a game that enables surprise, drama, comedy, and learning. The One Self could prevent all suffering by making error impossible; however, this would necessitate eliminating the avatars' free will. Although The One Self might be doing this elsewhere in the Universe, Earth is an example of a place where free will is the game being played. Perhaps, to The One Self, the avoidance of error is not as valuable as going through error, seeing what happens, and learning from it. In the latter case, understanding has been gained, and one is potentially capable of not only preventing oneself from making the same type of error again, but also of explaining to others so that they are prepared to avoid the same type of errors.

Then there is the type of suffering that one imposes on oneself through one's expectations and attachments. There is learning here too—learning to not be attached to outcomes, learning that one's happiness is not dependent on anything outside of oneself, learning acceptance.

And there's the suffering caused by fear—fear of losing something one has, fear of not getting something one wants, fear of some future imagined event, fear of what others might be thinking—most of which involve learning to live in the present moment and acceptance of those things and events not under one's own control.

In these latter cases, suffering is a choice one doesn't have to make. One can alleviate one's own suffering.

In the former cases, there is ultimately some level of choice in that one can give up attachment to one's physical body, understanding its impermanence and knowing one's soul or spirit will rejoin its Creator. And therein lies reason to rejoice.

The terrorist attacks on 9/11, and Hitler's holocaust, and so many other terrible sufferings we have all been through have convinced many people that there cannot be a God that cares about us. But their imaginations have not yet been stretched, to understand how all that changes, if death is not the end. With death as illusory as everything else in the multiverse, all that really matters is the value of experiencing and learning, and most of all, loving it all. In synch with the right attitude of The First Self, having the same cosmic perspective.

CHAPTER 8
EXPERIMENTATION FOR INDIVIDUALS

Experiment One[1]

Imagine for a moment the possibility that the voice or thoughts you hear in your head—which you think of as you—is sometimes actually not you. Imagine it is at least possible that sometimes this voice is God, living your experience through your body and mind. When

---
(1) Bill Harvey, *You Are The Universe: Imagine That* (New York: The Human Effectiveness Institute, 2014), pp. 244-246.

what you hear in your mind has the ring of truth and the power to change your life for the better, perhaps higher wisdom is being sent to you.

If this seems too hard to accept because of the connotations you have with the word "God", then call it something else. Call it the One Consciousness that has always existed. Or The One for short.

Or think of it as the intelligence of the Universe, which expresses itself through you and through a conceivably infinite number of other avatars, some like you, some like subatomic particles/wavicles, others of every kind, everything you see around you, all different expressions of the intelligence of the multiverse, of all that is.

Just suspend your disbelief for a moment. It might not be so easy to do, but it's not that hard once you get the hang of it. Play acting is a way to get into it. Act as if you are opening your mind wide, even if you don't believe yourself.

The obsession with 100% certainty, based on replicated laboratory experiments as the only way to gather useful insights into the nature of existence, traps and limits our epistemology, and prevents us from giving objective consideration to new ideas and new perspectives.

If it makes it easier to imagine, think of The One as not so different from a software program existing alone in spacetime, always having existed. A coruscating self-aware software program, with so much computing power and memory storage that the entire universe we know of can fit within one tiny corner.

Why would such a software program exist? How could something have always existed? These are mind-boggling questions. But so are the questions: Why would anything exist? How could something that has never existed before come into existence? Either way—the matter-based Big Bang model, or the always-existing software program—both are hard to imagine. The only difference is that we have become used to the entrenched Big Bang model. We have forgotten how to be amazed by it—for it has worn out its amazingness for us. But these are just feelings. And we should not base our view of reality merely on what we are used to, so much so that now it feels right.

Let's consider tuning our feelings back to a zero baseline in this regard—take a fresh look, start from scratch again, be a blank slate. Suspend belief and disbelief.

Just imagine it is at least conceivable that God—a wondrously extraordinary software program, a consciousness, an experiencer—is looking out your eyes.

He/She/It is having your experience. You are sensing this as *you* having that experience. How could you tell the difference?

One way you could tell the difference is if you find that you continue to experience *experiencing* after you die.[2] If this happens, then the experiencer has obviously not died. You could then say, okay, it's still me, I am just reincarnating or something. True, but you could also see it the other way: there might not be any separate "me", it might be God looking out through my experiencing-window, having my experience, making my decisions, talking to Himself/Herself in my experience-bubble.

All we are suggesting is that you imagine this is at least possible.

In fact, there is as yet no scientific evidence to say it is or isn't. The notion that our "separate consciousness" may not be separate after all, but rather God talking to Himself/Herself within us, is no stranger than the it-all-happened-by-accident Big Bang something-came-out-of-nothing explanation.

---

(2) If this theory is incorrect, you will not know it when you die because you will be dead and bereft of knowing.

If you can delicately balance your mind so that you don't buy into neither notion but just admit to yourself that either one (or something else) could be the true explanation of reality—then you will have reset your mind to zero base on this subject.

When this happens, there will come a point when you suddenly say to yourself, "Wow, it really could be true—anything could be happening here—I really have no idea—none of us do—and yet we all keep running around doing our thing as if it is a matter-based universe with nothing special, nothing meaningful going on in the background, no real point to existence."

Maybe that's the smart thing to do, and maybe it isn't. We are certainly not hedging our bets in any way. However, by basing everything on matter, we might be missing some of the best stuff life has to offer.

What then are we to do?

The sensible thing is to keep all options open and to explore all options more fully. Experiment carefully in all directions that pass the test of being positive, without any negative side effects to yourself or anyone else.

Experiment Two

This is the same experiment as Experiment One, with one difference. In the first experiment you opened your mind to the possibility that what you think of as your thoughts might actually be the Intelligence of the Universe, a central processor with infinite capacity to run an infinite number of avatars who in some cases may not be aware of the relationship — such as on Earth, at least in terms of the majority of its human inhabitants.

In this experiment you assume that at least some of the thoughts and feelings going through your mind and body are locally-produced, whereas others may come down the line from Central Processing.

This experiment also has positive side effects. It helps you learn how to listen up internally for what could be "pearls" among the more populous "swine"-like ideas and impulses in your mind. Our books *Mind Magic: Doorways into Higher Consciousness* and *You Are The Universe: Imagine That* expand on these ideas but we will focus on the experiment here.

This experiment reveals the essence of the Observer state: it is "always-on" metacognition. You are watching your every move, drilling down to the places where each thought and feeling arises. This has to be done and mastered in an alone state before it can be practiced and mastered in situations involving other people.

You might say at this point, well thanks mate for the interesting and thought-provoking ideas, but this is where I get off. I simply haven't got the time to study myself like a lab specimen, even if it's only for twenty minutes a day.

Ah, but wait. We're simply suggesting twenty minutes a day of meditation can be beneficial—an idea which is becoming more mainstream every day. The experiment described above is what meditation is. We're just demystifying it here, making it more operational.

In the process of the experiment (which can start a whole new stage in life that never ends), focus on the difference in level of the individual thoughts and feelings. Some will seem to come from high places, some from low—you have probably already experienced that all your life. But if you listen to the very whispery voices, when the ego has gone temporarily silent with its always-on blaring propaganda campaign, you will earn inspirations from upstairs (if the theory is directionally correct).

Experiment Three

By this stage you will already be working out your avatar relationship with the Intelligence of the Universe. This experiment in fact is about conversing with the Intelligence, just in case it is there.

Why talk to yourself, ruminating about your thoughts, when you can talk to God, the Original Self of all of us, who is actually inside you in the silent spaces?

Listen patiently and attentively in the silences, for there is where the response may come, sometimes days later.

It may help to keep a journal. And it is the scientific way.

What to talk to one's highest Self about? The instinct is not to be too petty.

You can talk freely because you are talking to yourself. It's just that your Self happens to be the supreme Intelligence of the multiverse, just like in Marvel Comics (though even *they* haven't gone that far yet). This is the hypothesis and the reason for the experiments. We are hoping that each individual through these experiments can learn what we see to be the truth, i.e., that we are all One. It can be talked about on the outside, but it can only be learned to be Truth on the inside, one person at a time, each achieving It with help from above (if the theory is directionally correct).

Experiment Four

Keep track of synchronicities in your journal. Why did the (hypothetical) Intelligence of the Universe make you hear that line in the audio at the moment when it coincided with what you were just thinking—"now what was it that I was thinking..."—and is likely to be a good thing to find out. The Central Office wants to either encourage or discourage you in certain courses of action; trust your true feelings to get the valence of the suggestion—this is the experimental task.

Tracking synchronicities in your journal will likely provide you with sufficient if not indisputable evidence to realize that there really is an accessible higher guidance system. What better solution for the world's nightmares, if everyone realized this and tuned in to this higher guidance system? These are testable hypotheses, and they are worth testing.

Experiment Five

Also use your journal to keep track of situations where you felt as if someone was reading your mind, or vice versa. Include details on the meaning of the situation and any predictive inference. Feel free to go back and discuss such situations with the people involved and include this in your journal as well.

Whenever you can get away with it, speak up in the moment to say something like, "You must be reading my mind," or to ask questions when appropriate so as to collect more information on the possible telepathy event.

Experiment Six

Another journal usage is to track your hunches, and to compare the hunch with the outcome. This starts with entering the date and time, the event description, the hunch description, and any thoughts you have about the hunch.

You will be able to look back and see how well each hunch performs, and make journal entries indicating how pragmatically helpful the hunch turned out to be.

Hypothesis: You will be surprised to find that in a large proportion of cases, your hunches were correct, though you may or may not have taken advantage of them.

φ

Thank you for reading this short book and engaging with this other part of the total consciousness (me). And for making the time to see if the experiments might have life changing effects on you.

Enjoy the journey!

Love, Bill

ABOUT THE AUTHOR

Emmy® Award winner and media research industry leader, Bill Harvey has been lauded as a visionary and technological pioneer of the changing mediascape over the last 35 years.

With an imaginative, unorthodox mind for research, innovation and invention, Bill started his career in the media business. He predicted today's media reality with his *MediaWorld 1990* report to the industry and in his widely-read *Media Science Newsletter*, and he invented media research tools and measurement systems, including some now written into FCC regulations.

Bill became a leader in the field as media morphed into being more interactive, putting the viewer in charge. In 2022, Bill received an Emmy® Award for the pioneering development of technology which protects privacy while collecting massive data on media usage at an aggregate level. In practical terms, this invention helps diverse audiences with a wide range of interests to get all the types of programs they enjoy the most.

Considered the dean of living media scientists, Bill in 2008 received the Advertising Research Foundation's *Great Mind Award*, and in 2014 he became the first recipient of the ARF's *Erwin Ephron Demystification Award*. He holds four issued US patents and has consulted for over a hundred Fortune 500 companies.

Working with neuroscientists, academics and industry on learning about the cognitive mechanisms connecting motivations and attention, Bill's company Research Measurement Technologies distilled a motivational taxonomy predictive of sales and audience behavior from every word in the English language in a series of large-scale experiments. This "memome" offers potential to the progress of psychology and to the next generation of AI Large Language Models.

Bill first experienced the Zone—that space where innovative and successful ideas and actions flow out of you effortlessly—as a young child. The son of legendary orchestra leader/emcee Ned Harvey and former Ziegfeld Follies showgirl Sandra Harvey, Bill started performing on stage at age four, dancing with showgirls and exchanging lines with comedic greats like Jack E. Leonard. He liked this feeling of being "on" and wanted to learn how to be "on" more often—and so began his lifelong quest to understand how to bring on higher states of consciousness and to help others do the same.

Earning his degree in philosophy, the first school subject he ever loved, Bill founded the Human Effectiveness Institute, with the goal of sharing the consciousness techniques he had learned and developed. Bill's ideas didn't all come from the inside, as he was inspired by something that Milton Berle once told him: "Always steal from the best, kid." His ideas were further inspired by Alan Watts, Buddhism, Zen, and by his parents and adopted older brother, the multitalented Bill Heyer, second trumpet in Ned's band.

Unique and ahead of its time, Bill's first book, *Mind Magic: The Science of Microcosmology*, met with rave reviews in the late 70s and was praised by thousands of readers whose lives, they say, were changed by it. Fans included John Lennon, Ram Dass, Norman Cousins, Daniel Goleman, and Jimmy Carter. The book has been used at 34 universities including NYU, UCLA and West Point, and by numerous organizations, and is now available in its 6th edition, *Mind Magic: Doorways into Higher Consciousness*.

In his second book, *You Are The Universe: Imagine That*, Bill speculates about the true nature of reality, in which all that exists is a single divine consciousness made of information. In this view, religion is not at odds with science.

Bill turned his theory into fiction, and conceived an epic series of novels entitled *Agents of Cosmic Intelligence*. *The First Son* was the first to be released, Episode 2 in the series, covering the ancient world and its great prophets—some of whom Bill casts as cosmically-inspired Agents.

The Message, Episode 11 in the Agents series, and *Pandemonium: Live to All Devices*, Episodes 12 and 13 in the series, were both published in 2022. *The Great Being*, Episode 1 in the series, is scheduled for publication in Fall 2023. Descriptions of these titles can be found in the following pages.

Bill lives with his wife Lalita and their cat Zohreh in New York's beautiful Hudson Valley. He has a daughter Nicole; four grandchildren, Nicholas, Gabrielle, Jessica and Alexander; and one great-grandchild, Zara.

FURTHER READING

You Are The Universe: Imagine That. A theory of what the universe is, reconciling science with religion. This book shows you a way of looking at reality that changes the way you look at yourself. You see the highest use of your life. Your own life becomes more exciting and inspiring. You have new understanding of individuals and your love flows easily.

Bill shows that the precepts of the major religions find support in this theory, but in the modern age have been obstructed from communicating their true wisdom as a result of outdated language, spiritual materialism, and the forgotten ability to reach into one's own essence feelings. Unlike the unified field theories of Albert Einstein and modern unified field theorists, Bill's picture of the universe can be visualized by non-mathematicians. Anyone can test this theory by means of objective scientific experimentation.

Sheds light on humankind's ancient burning questions, which boil down to… "What's going on here?"
—Peter Sorensen, England

Mind Magic: Doorways into Higher Consciousness is a guide to discovering who you really are beneath all the influences of other people and things in your life. What do you really want to do with your life? What will make you most effective and bring out your creativity in meeting life's challenges?

Reading **Mind Magic** is a unique experience. You glide along effortlessly, stimulated in light, sometimes humorous and often unexpected ways. It is designed to evoke your ideas, and get you thinking and acting in new ways.

There is no fixed formula for how to use **Mind Magic**. Some people enjoy opening it to random pages, finding they get just what they need at that moment. Others read it all the way through, going back to it again and again.

Over the years since its original publication in 1976, thousands of readers have written letters sharing their experience of **Mind Magic**.

If everyone were to read just this one book, the improvement in social and personal consciousness would be astounding. It's a marvelous inspiration.
—Lynn S., Indiana

AGENTS OF COSMIC INTELLIGENCE
the epic adventure chronicle of the universe

Agents of Cosmic Intelligence is a sweeping saga that unfolds forward and backward in time to become a story of the universe, how it might have "started" and where it might be "going."

Books in the Series

The Great Being
Episode 1 | 200,000 BC—3068 BC | Fall 2023

There is a Great Rebellion going on throughout the Universe, all of which is a single Mind at play. Two Agents of Cosmic Intelligence, Melchizedek and Layla, are dispatched to infiltrate the Rebels on Earth. However, the Rebels have interfered with Earth evolution, so the human brains that the Agents step into repress their knowledge of who they really are.

The First Son
Episode 2 | 3067 BC—27 AD | October 2018

Rebel groups form nation-states to continue the endless wars they have propagated to make Earth people the toughest fighters in the Universe, to eventually storm the gates of Heaven and take over the Multiverse entirely. The First Son and the Agents quietly build up the character of Earth humans by incarnating as great Teachers, beginning and spreading that tradition across the planet.

I love the way Harvey used science fiction to express the deep truth of the unity of all things. and the common theme of all wisdom literature. An engaging story line and beautiful prose. —Jim Spaeth, NY

Action-packed sci-fi adventure of a supremely mind-blowing kind. A real page-turner, always surprising.
—Dennis, Amazon Review

The Message
Episode 11 | Ca. 2035 | December 2022

The five Agents have all fallen asleep to their true identities. Four of them lead the U.S. Army's top-secret psychic Theta Force. Nastassia is in a rival unit in Russia. The Agents suddenly become uplifted to the highest state of consciousness they ever remember experiencing, as a result of a Message apparently from outer space that is heard by every psychic on Earth, though each one hears something slightly different.

Bold, brisk, conspiratorial psychic thriller imagines humanity's secret history. Harvey takes wild narrative risks readers will not see coming… it's the ideas that drive this [Agents of Cosmic Intelligence] *series: Harvey spins a secret history of all of us, urging us to be more.*

—BookLife Reviews by *Publishers Weekly*

Pandemonium:
Live to All Devices
Episodes 12 & 13 | Ca. 2037 | June 2022

In this fast-moving thriller, a heady amalgam of hidden war, psychics, Nazis, aliens, artificial intelligence, virtual reality, and transcendental love takes place against a backdrop wherein the latest media/technology revolution triggers sudden unprecedented changes in world politics.

Harvey continually upends reader expectations... daring to go bigger and stranger, the in-the-moment suspense connected to the mind-blowingly cosmic. Pandemonium *gets wilder as it goes, with international romance and a savvy sense of how media shapes minds, nations, and history.*
 —BookLife Reviews by Publishers Weekly

Pandemonium *is a masterful piece of cautionary fiction that will likely ring with relevance for years to come.* —Independent Review of Books

PRAISE FOR BILL HARVEY'S BOOKS

MIND MAGIC

Highly recommended... will loosen your moorings and open you to creative vistas.
—Dr, Daniel Goleman,
author of Emotional Intelligence

MIND MAGIC is a delight. Sets forth with neat precision just how to do it (think).
—Ram Dass, author of Be Here Now

What sets your book apart from all others in this field... is that it is a rare combination of frontier knowledge, wisdom, and plain old fashioned warmth... in your debt for the insights it provides.
—Norman Cousins, founding editor of Saturday Review

YOU ARE THE UNIVERSE

How great it is that your book came along at this time. It's just what I asked for... a way of understanding the true nature of my self.
—Richard Fusco, Music Industry Executive

Bill Harvey's writing is courageous... not just with the effort he puts into his words but with the ideas he asks his readers to explore, which can benefit their lives.
—Bob DeSena, Marketing Executive

THE FIRST SON

In this stimulating novel, Bill Harvey takes what Aldous Huxley called The Perennial Philosophy to the next level of detail. Harvey offers an imaginative approach to the underlying dynamics of how this philosophy might have gradually permeated human thought from the Hebrew Bible, though the Mahabharata, Buddhist Sutras, the I Ching, Plato's Academy and even Christianity. Harvey accomplishes this task by integrating modern terms such as artificial intelligence... with classical concepts like avatars... bold vision.
—Mike Hess, Polymath Scientist

I love the way Harvey used science fiction to express the deep truth of the unity of all things. and the common theme of all wisdom literature. I also appreciated the use of the "rebel" metaphor to explain the perversion of that wisdom. All expressed through recognizable personalities, an engaging story line and beautiful prose.
—Jim Spaeth, Sequent Partners

THE MESSAGE

Whether for its futuristic social experiments in transformation, its riveting action-packed world, or the changes characters experience in the redefinition of their perceptions and purposes—the story is complex, inviting, and hard to put down.
—Midwest Book Review

Bold, brisk, conspiratorial psychic thriller imagines humanity's secret history. Globe-crossing thriller of expanded consciousness... The story this time turns on a surprise transmission of uncertain origin received by all of Earth's psychics: an androgynous voice saying, among other things, "Each of us is God... Harvey takes wild narrative risks [that] readers will not see coming. Harvey's inventiveness extends to cosmic alt-history, millennia-spanning conspiracies, futurology, untapped human potential, memorable tech-thriller action... But it's the ideas that drive this [Agents of Cosmic Intelligence] *series: Harvey spins a secret history of all of us, urging us to be more.*
—BookLife by Publishers Weekly

It's a pleasure to suspend your disbelief in the face of this semi-dystopian premise with a wide cast of colorful characters. Harvey isn't afraid to dig deep into a particular type of tech, an ideology, or a military procedure, and that storytelling flair for flowing in and out of detail keeps the rhythm enticing and the read addictive. For an inventive work of visionary sci-fi with an allegorical tinge of social commentary, The Message *is a riveting addition to Harvey's enigmatic* Agents of Cosmic Intelligence *series.*
—Independent Review of Books

Thoughtful and provocative... [Isaac] Asimov would be proud and [Neal] Stephenson will be overjoyed to have Harvey as a contemporary.
—BookTrib

PANDEMONIUM

Harvey continually upends reader expectations... daring to go bigger and stranger, the in-the-moment suspense connected to the mind-blowingly cosmic. Pandemonium *gets wilder as it goes, with international romance and a savvy sense of how media shapes minds, nations, and history.*
—BookLife by Publishers Weekly

A science fiction spy thriller with deep metaphysical undertones where the future of humankind is at stake. Fueled by brilliantly insightful scientific speculation... meticulously described science fiction backdrop... full of big ideas... mind-bending and thought-provoking.
—BlueInk Review

The depth of psychological examination of individuals, organizations, country leaders, and worldviews makes for an outstanding enhancement of the usual sci-fi attention to action and confrontation... Gives Pandemonium *a transformative feel and quality that makes it as much an intellectual pursuit as an action-packed adventure. Bill Harvey leads readers on a satisfyingly unpredictable romp through a future world in which special interests have humanity's future in thrall... Controlling outcomes and confronting evil has never felt so multifaceted and thoroughly engrossing.*
—Midwest Book Review

Plentiful and accurate projections for the future... each element is conceivable both in terms of technology and the sociology behind the ways in which it has evolved. The inspiration behind these technological advancements are all born from trends we can see in society today, and will have readers thinking long after they close the book. With the scope of Asimov and the prescience of Bradbury, Bill Harvey makes the greats proud.

—BookTrib

www.ingramcontent.com/pod-product-compliance
Lightning Source LLC
Chambersburg PA
CBHW031427290426
44110CB00011B/559